NIAGARA

NIAGARA

Peter Fowler

Toronto
OXFORD UNIVERSITY PRESS
1981

Plate 42 is reproduced by courtesy of Robert G. Ragsdale, ARPS
and the Shaw Festival, Niagara-on-the-Lake

Designed by Fortunato Aglialoro

ISBN 0-19-540

1 2 3 4—4 3 2 1

Printed in Hong Kong by
EVERBEST PRINTING COMPANY LIMITED

THE FIRST DESCRIPTION OF NIAGARA, 1678

I could not conceive how it could be that four great lakes, the least of which is four hundred leagues in circuit, and which empty into one another, which all come at last massed at this great fall, do not inundate this great part of America. What is more surprising in this is that from the mouth of Lake Erie to this great fall the land appears almost all smooth and level. You can scarcely perceive that one part is higher than another, and this for the space of six leagues. It is only the surface of the water, the current of which is very rapid, that makes it noticeable .... And yet the discharge of so much water, coming from these freshwater seas, centres at this spot and thus plunges down more than six hundred feet, falling as into an abyss which we could not behold without a shudder. The two great sheets of water, which are on the two sides of the sloping island that is in the middle, fall down without noise and without violence, and glide in this manner without din; but when this great mass of water reaches the bottom, then there is a noise and a roaring greater than thunder.

*Moreover the spray of the water is so great that it forms a kind of clouds above this abyss, and these are seen even at the time when the sun is shining brightest at midday .... It is into this gulf, then, that all these waters fall with an impetuosity that can be imagined in so high a fall, so prodigious, for its horrible mass of water. There are formed those thunders, those roarings, those fearful bounds and seethings, with that perpetual cloud rising above the cedars and spruces, that are seen on the projecting island, already mentioned.**

*Father Jean-Louis Hennepin (1640-c.1705), a Recollet priest who viewed Niagara Falls in 1678. From *Nouvelle découverte d'un très grand pays situé dans l'Amérique*...(1697) in the Appendix to *A Description of Louisiana by Father Hennepin* (1880), translated and edited by John Gilmary Shea.

NIAGARA 1871

By Henry James

Though hereabouts so much is great, distances are small, and a ramble of two or three hours enables you to gaze hither and thither from a dozen standpoints. The one you are likely to choose first is that on the Canada cliff, a little way above the suspension bridge. That great Fall faces you, enshrined in its own surging incense. The common feeling just here, I believe, is one of disappointment at its want of height; the whole thing appears to many people somewhat smaller than its fame. My own sense, I confess, was absolutely gratified from the first; and, indeed, I was not struck with anything being tall or short but with everything being perfect. You are, moreover, at some distance, and you feel that with the lessening interval you will not be cheated of your chance to be dizzied with mere dimensions. Already you see the world-famous green, baffling painters, baffling poets, shining on the lip of the precipice; the more so, of course, for the clouds of silver and snow into which it speedily resolves itself. The whole picture before you is admirably simple. The Horseshoe glares and boils and smokes from the centre to the right, drumming itself into powder and thunder; in the centre the dark pedestal of Goat Island divides the double flood; to the left booms in vaporous dimness the minor battery of the American Fall; while on a level with the eye, above the still crest of either cataract, appear the white faces of the hithermost rapids. The circle of weltering froth at the base of the Horseshoe, emerging from the dead white vapours—absolute white, as moonless midnight is absolute black—which muffle impenetrably the crash of the river upon the lower bed, melts slowly into the darker shades of green. It seems in itself a drama of thrilling interest, this blanched survival and recovery of the stream. It stretches away like a tired swimmer, struggling from the snowy scum and the silver drift, and passing slowly from an eddying foam-sheet, touched with green lights, to a cold verd-antique, streaked and marbled with trails and wild arabesques of foam. This is the beginning of that air of recent distress which marks the river as you meet it at the lake. It shifts along, tremendously conscious, relieved, disengaged, knowing the worst is over, with its dignity injured but its volume undiminished, the most stately, the least turbid of torrents. Its movement, its sweep and stride, are as admirable as its colour, but as little as its colour to be made a matter of words. These things are but part of a spectacle in which nothing is imperfect. As you draw nearer and nearer on the Canada cliff to the right arm of the Horseshoe, the mass begins in all conscience to be large enough. You are able at last to stand on the very verge of the shelf from which the leap is taken, bathing your boot toes, if you like, in the side-ooze of the glassy curve. I may say, in parenthesis, that the importunities one suffers here, amid the central din of the cataract, from hackmen and photographers and vendors of gimcracks, are simply hideous and infamous. The road is lined with little drinking-shops and warehouses, and from these retreats their occupants dart forth upon the hapless traveller with their competitive attractions. You purchase release at last by the fury of your indifference, and stand there gazing your fill at the most beautiful object in the world.

The perfect taste of it is the great characteristic. It is not in the least monstrous; it is thoroughly artistic and, as the phrase is, thought out. In the matter of line it beats Michelangelo. One may seem at first to say the least, but the careful observer will admit that one says the most in saying that it *pleases*—pleases even a spectator who was not ashamed to write the other day that he didn't care for cataracts.

Portraits of Places

1 The Peace Bridge between Fort Erie, Ont., and Buffalo, N.Y.

NO TAX
LOWEST PRICES
50 YEARS
BUFFALO
PEACE BRIDGE
CANADA
Drive
Safely
the Peace Bridge

2 Niagara has been a honeymoon attraction since the early nineteenth century.
3 Garden of the Horticultural School

4 Ball's Falls Conservation Area

5 Queen Victoria Park

6 Queen Victoria Park Boulevard
7 St Catharines

8 Lower Niagara River towards Queenston, with the Spanish Aero Car

9 Horseshoe Falls and American Falls

CAVE OF THE WINDS TRIP
MM
MAID OF THE MIST

10 American Falls with the 'Cave of the Winds' Walk and the *Maid of the Mist*
11 Horseshoe Falls with a helicopter ride

12 Skylon Tower
13 Table Rock lookout platform, at the Horseshoe Falls

14 Floral Clock outside the Hydro Hall of Memory, Queenston
15 The brink of the Horseshoe Falls

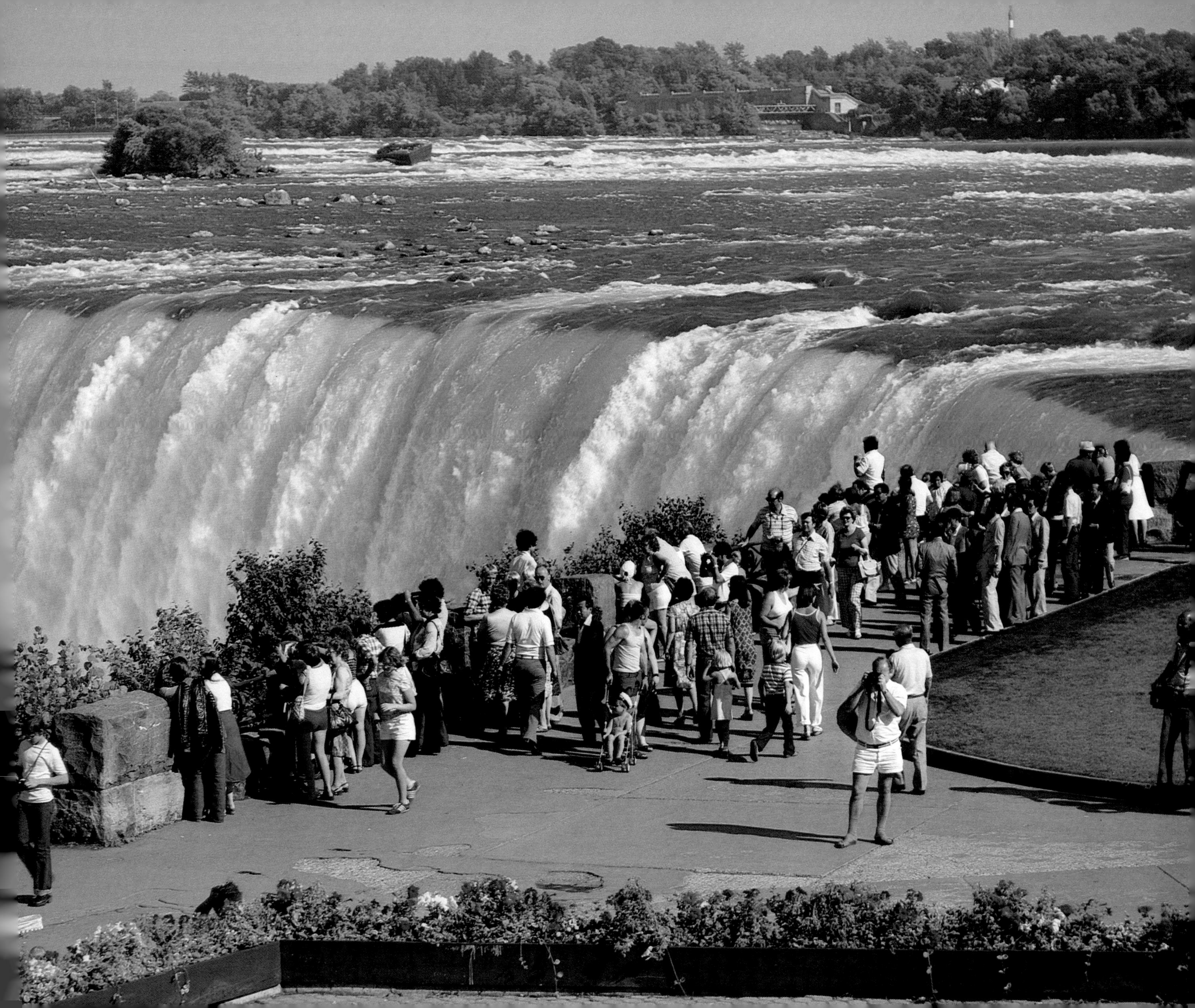

16
&
17 Maple Leaf Village

18 Tivoli Gardens

19 Crystal Beach

20 Port Dalhousie, St Catharines
21 Marineland

22 & 23 Marineland

24 Fort George, Niagara-on-the-Lake
25 Niagara Grape Festival Parade,
St Catharines

26 & 27 Fort Niagara, Youngstown, N.Y.

SKYLON INTERNATIONAL MARATHON
START

28 Start of the Skylon International Marathon to Niagara Falls, Ont., from Buffalo, N.Y.
29 Port Colborne

30 Niagara Falls
31 Niagara Grape Festival Parade, St Catharines

UNIQUES

32 Fort George, Niagara-on-the-Lake
33 Niagara Falls a & b

1981
Niagara Falls,Canada
Niagara Falls,Canada
Niagara Falls
Canada
NIAGARA FALLS, CANADA
Niagara Falls,Canada

a

NIAGARA FALLS, CANADA

b

34 Louis Tussaud's Wax Museum,
Niagara Falls

35 Winona Peach Festival

M. SACZKOWSKI
CANADA NO. 1
6 QUARTS - 6 PINTES
6.819 LITRES
CANADA NO. 1
6 QUARTS 6 PINTES
6.819 LITRES
BIJAKOWSKI ORCHARDS
CANADA NO 1
6.819 LITRES

36 Roadside fruit-stand, Vineland
37 Niagara Grape Festival, St Catharines

38 Jordan Valley Flea Market
39 Fruit-stand, Grimsby

40 Letchworth House, John St,
Niagara-on-the-Lake, built 1829

41 Niagara Falls

42 Shaw Festival, Niagara-on-the-Lake
43 Prince of Wales Hotel, Niagara-on-the-Lake

PRINCE OF WALES HOTEL
ESTABLISHED

a

b

c

d

44 a The Niagara Apothecary, Niagara-on-the-Lake
44b Battlefield House, Stoney Creek, built in 1796 by James and Mary Gages as a homestead
44c Stewart-McLeod House, Prideaux St, Niagara-on-the-Lake, built c.1813 by the family of Alexander Stewart
44d Youngstown, N.Y.
45 Richardson-Kylie House, Queen St, Niagara-on-the-Lake, built c.1832 by Charles Richardson

46 The house of William Lyon Mackenzie (1795-1861), Queenston, where he founded *The Colonial Advocate*, in 1824

47 Breakenridge-Hawley House, corner of Mississauga St and William St, Niagara-on-the-Lake, built c.1818 by John Breakenridge

48 Fort Niagara, Youngstown, N.Y., looking across the Niagara River to Niagara-on-the-Lake
49 Court House Theatre, Niagara-on-the-Lake

COURT HOUSE THEATRE

50 Niagara Cruises, Niagara-on-the-Lake
51 Royal Canadian Henley Regatta, St Catharines

2

52 Smithville
53 Beamsville

54 Grimsby
55 St Catharines

MOUNTAINMILLS
DECEW FALLS
GRISTING & CHOPPING
W.MORNINGSTAR
CASH WHEAT
FLOUR & FEED ALWAYS ON HAND

56 Wray Patterson's General Store, Effingham
57 Dufferin Islands Park, Niagara Falls

58 Grimsby
59 Winona

60 Beamsville
61 Virgil

62 St David's
63 Fonthill

64 Smithville

65 Stevenville

66 Brock Monument, Queenston Heights Park, Queenston. It marks the burial place of Major-General Sir Isaac Brock (1769-1812).
67 Niagara Falls

68 Grimsby
69 Pelham

262

70 St Catharines
71 Queenston

72 Queenston
73 Virgil

74 Niagara Falls
75 Jordan Harbour

76 Niagara Falls, N.Y., looking towards Canadian side
77 American Falls in winter

78 The home of Laura Secord (1775-1868), Queenston
79 Grimsby

STOKES SEED FARMS

80 St Catharines
81 Smithville

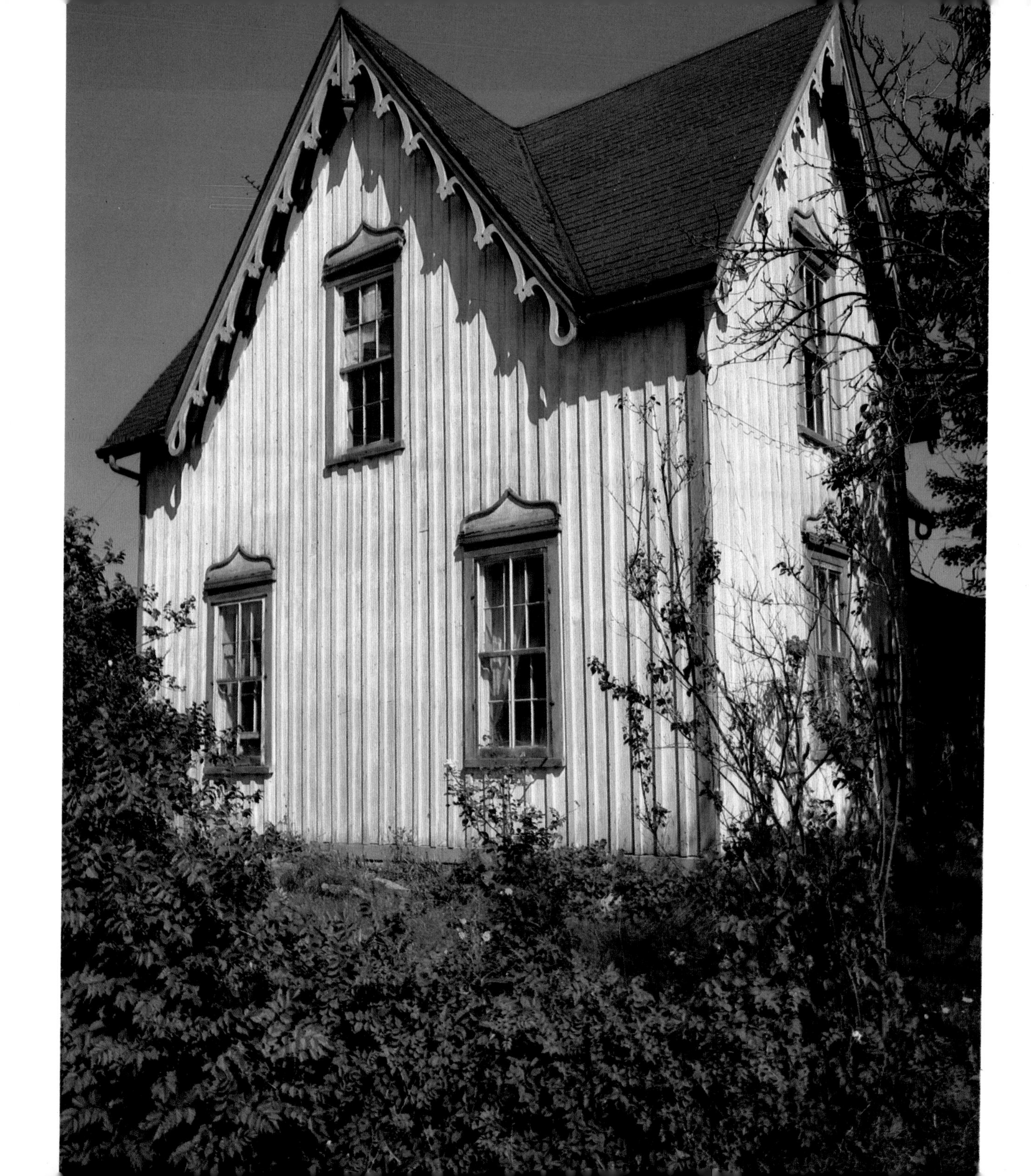

82 Welland Canal
83 Niagara River at Youngstown, N.Y.

84 Skylon Tower and Horseshoe Falls
85 *Maid of the Mist* vessels, with the Rainbow Bridge, Niagara Falls

MAID OF THE MIST
MAID OF THE MIST
MAID OF THE

86 Panasonic Tower, Niagara Falls
87 American Falls
88 Horseshoe Falls